Looking Back

PORCUPINE
GOLDFIELDS
ONTARIO

This photograph of an unidentified miner was taken at the Dome Mine in around 1915. At that time, miners were still using candles for light underground; his is tucked into his cloth cap. He is covered in wax drippings.

Looking Back

PORCUPINE
GOLDFIELDS
ONTARIO

Karen Bachmann

Looking Back Press

Vanwell Publishing acknowledges the financial support of the Government of Canada through the Book Publishing Industry Development Program for our publishing activities.

Published by Looking Back Press
An Imprint of Vanwell Publishing Limited
1 Northrup Crescent, P.O. Box 2131
St. Catharines, ON L2R 7S2
For all general information contact Looking Back Press at:
Telephone 905-937-3100 ext. 835
Fax 905-937-1760
E-Mail vanessa.kooter@vanwell.com

For customer service and orders:
Toll-free 1-800-661-6136

Printed in Canada

National Library of Canada Cataloguing in Publication

Bachmann, Karen, 1963-
      Porcupine Goldfields, Ontario / Karen Bachmann.

(Looking back)
Includes bibliographical references.
ISBN 1-55068-922-3

      1. Gold mines and mining—Ontario—Timmins Region—History. 2. Gold mines and mining—Ontario—Timmins Region—History—Pictorial works. 3. Timmins Region (Ont)—History. I. Title. II. Series: Looking back (St. Catharines, Ont.)

FC3099.P5638B32 2003   338.2′741′09713142   C2003-902716-3
F3099.P5638B32 2003

H. Peters took this photograph of prospectors at the mouth of the Porcupine Trail in 1909. They were headed for the Porcupine Goldfields.

Contents

Acknowledgements

This book has given me the opportunity to feature images from the photograph collection of the Timmins Museum: National Exhibition Center. This collection provides a glimpse into the development of the Porcupine region, and I am indebted to the photographers and collectors who have helped create it. I enjoyed the challenge of identifying some of the photographs and making the selections for this book.

I refer to the area as the Porcupine Goldfields (the official name at the time of the first gold rush), the Porcupine Camp (which included the seven townsites around the mines), and the communities by their individual names; all are roughly representative of the region now known as Timmins.

I am also thankful to the staff of the site who helped make my job easier, and to friends and family for their encouraging support and words of advice. I would like to thank Vanessa Kooter, my editor, who guided me through this process (I am sure at the expense of her sanity!). I hope to be able to continue with the exploration of the collections, and to feature them in a second volume.

A Note on Photographs

William Henry Peters, a prominent pioneer of the Porcupine, took most of the photographs featured in the book. Mr. Peters recorded the initial development of the Porcupine Goldfields and provides us with a wonderful glimpse of the beginnings of our community. From the Fire of 1911, to the sale of land around current-day Schumacher, Mr. Peters was at the ready to photograph the early settlers.

William Henry Peters was born in Bobcaygeon in 1863, and lived in a number of communities across Ontario, including Little Current on Manitoulin Island.

As a photographer, he travelled extensively throughout Canada and the United States, photographing lumber camps, outdoor scenes, and mining operations. He was lured back to New Ontario, and the Porcupine Camp in May of 1911. Excited by what he saw, he proceeded back to Cochrane to gather his equipment, and made it back just in time to see and record the destruction of the area in the Great Porcupine Fire of July 11th, 1911. He photographed the many scenes of heroism and destruction, and those photographs provided him with a comfortable stake.

He moved to Timmins in September of 1911, where he built a log building and summoned his family. He continued to photograph the people and events of the community, including the flood of 1913, and unwittingly provided us with our best views of the creation of Timmins, South Porcupine, Schumacher, and Porcupine. "Harry" Peters was the town's first postmaster after Timmins incorporated in 1912. He also served on Council, was an active member of the Oddfellows, the local Board of Trade, and the Masons. Mr. Peters died in 1922, while undergoing treatment for "hardening of the spine" in Owen Sound, Ontario.

The museum is indebted to the family of Mr. Peters who generously donated most of his work to our collection. This gift will be preserved so that future generations can reflect on Mr. Peters' unique portrait of the Porcupine.

R. Tomkinson, a photographer from South Porcupine, and Charles See, a businessman from the camp, took other photographs included in the museum's collection and in this book. The newly acquired H.P. Davies Collection yielded never before seen shots of the early days in the Porcupine

Introduction

If you drive about seven hundred kilometers north of Toronto, you eventually arrive at the junction of highways 11 and 101. Highway 11 will take you to Cochrane and north to the Trans-Canada Highway. If you choose to take highway 101, which will lead to Chapleau and other rugged northern locales, you will find yourself on a road lined with Jack pines, and muskeg. You will breeze through small towns like Matheson and Hoyle, tiny communities that have sprung up along the railway and old fur trade routes. You will come upon working smelters and abandoned head frames, old rickety shacks and modest modern homes, acres of nothing and crowded city blocks. You could speed through Porcupine, South Porcupine, Schumacher, and Timmins without ever knowing that you were driving through one of the world's richest gold mining communities, an area whose gold rush equaled or surpassed that of the Klondike. You wouldn't know that Timmins, this little city currently known as the "hard-up" home town of an internationally renowned singer, has a history that extends well beyond her success and the modest mentions on the Ontario Blue History Plaques that dot the area.

Timmins, the largest city in the Porcupine Goldfields, is a community challenged by geographic location, geological formations, and a winter that sometimes won't quit. It sits on the Canadian Shield, a terrain that is pockmarked by small lakes and streams, remnants of the last Ice Age. If you have an interest in geology and the intricacies of the science, you can see almost all of the basics in Timmins. Eskers, exposed shield rock, pillow lava—all are in evidence. It also has some of the most exciting exploration fields in the country. When the gold market is up, prospectors still arrive in droves looking for gold and base metals. When times are poor, prospectors and speculators still congregate in local hotels and talk about the next big play, which is always just within their grasp, with the right investment, of course. Timmins is, contrary to current media depictions, a city that simply won't roll over and die—it plays up an easy-going "here today, gone tomorrow" philosophy that characterizes cities and towns in northern Ontario. It remains true to its roots, proud of a mining heritage that is notorious for keeping a secret. The story of the early Porcupine Goldfields (Timmins, Schumacher, South Porcupine, and Porcupine) is filled with a northern mythology that still permeates the community today. It is the legends and facts of the first ten years of the camp that will be the focus of this book; hopefully it will help to shed some light on the north's "best kept secret".

One
The First Steps

While 1909 is generally regarded as the year the Porcupine Goldfields were discovered, evidence exists that the area had been inhabited for over 5000 years. The Night Hawk Lake area near Timmins was home to a people known as the Shield Archaic Culture; artifacts found around the lake tell us the group was well established along the waterways in the region. They lived a life very similar to that of the Cree, Ojibwa, and Algonquians during the later period; it is believed that the Shield Archaic were probably the ancestors of these groups.

Night Hawk Lake remains an important area in the history of the region; it is believed that in the late seventeenth century, le Sieur d'Iberville, a French explorer sent to find a route from the Ottawa region to Hudson's Bay explored an island in the middle of the lake and noticed an abundance of precious metals. While measurably interested in the possibility of finding gold, le Sieur was obsessed with the idea of capitalizing on the fur trade and proceeded northward where he attacked the English fur trade post at Moose Factory. He remained there as a trader until the post and the region were ceded to the English at the treaty of Utrecht in 1713. Naturally, the fur trade gradually moved south out of James Bay and forts were established at Night Hawk and Frederick House.

Because of this, traffic in the area increased dramatically. While some half-hearted attempts were made to uncover more of the showings along the Night Hawk system, no attention was paid to the few diehards who had small mining operations on the island.

It wasn't until the nineteenth century that Ontario turned its attention to Rupert's Land and started the serious business of surveying the land. Alexander Niven was sent up in 1896 to run the survey line from north to south through the territory; it would become an extension of the boundary between the Algoma and Nipissing districts. He ran another baseline in 1900 that ran west; along with Niven's Meridian, these lines became the basis for all surveys in this part of Ontario.

Along with government surveyors came the government geologists. E.M. Burwash and W.A. Parks roamed the area recording gold showings in exposed quartz veins from the Algoma-Nipissing boundary to the portage route on the Mattagami to Whitney Township. But generally, it was believed that Ontario gold would never run deep. While their reports did not stir up much excitement at the time, some prospectors nevertheless made their way to the region.

The great nickel belt around Sudbury was soon "under the pick," and Charles Campsall was sent to the Algoma region to look for iron ore. On his way back, he visited the Night Hawk Lake area and, since winter was fast approaching, grabbed a few samples from the outcropping of quartz and hurried back to Sault Ste. Marie. The samples he tested ran rich, but because of the remote location and the distance to the railway, production was seen as too costly a venture. The Porcupine would have to wait.

In 1904, Reuben D'Aigle, a prospector out of New Brunswick worked the same area as Campsall but found nothing. He returned the next year with a team and sank a shaft on Hollinger Hill that produced absolutely nothing. He gave up before the end of the season and went home to grubstake another expedition. Others staked the Night Hawk area and small mining operations were started in 1907 and 1908. The Night Hawk Peninsular Mine was the most successful, producing a handsome sum for the owners, during a somewhat sporadic operating career. But as always, it appeared that things were better elsewhere, and the Porcupine was obscured by the silver rush at Cobalt and the rumours of gold in Gowganda and Kirkland Lake. It would have to wait yet again for the big rush.

The Cree and Ojibwa communities of the area had settlements along the Frederick House River and throughout the waterway system to James Bay. They played an important role in the fur trade as they worked with the trading posts in the region. This family lived in the Porcupine at the time of the gold discoveries.

A tikanagan, or cradleboard was used to easily carry children on the trail and in town. A large piece of hide or cloth was fastened to the board and laced up around the little one. Blankets, cloth,. or sphagnum moss and cattail down were stuffed around the infant for additional warmth. The board was carried on the mother's back; a leather strap attached to the top of the board passed around her forehead.

In 1889, the British Parliament confirmed Ontario's claim to a vast stretch of land we now know as Northern Ontario. With that declaration came a mad rush to discover the lands in the region. The Ontario government sent surveyor Alexander Niven to run a survey line from north to south through this territory. A baseline was drawn in 1898, while a second baseline running west of milepost 198 came two years later. The two young men on this team were part of the Gowganda to Porcupine survey crew.

These two men are using a survey transit to make their marks. They would have also used a range pole to help with the sight, as a plumb bob was difficult to see in the dense bush. Gunter's chains (a measuring tape consisting of 100 links, 66 feet total length) were not used in the bush, as they were too bulky and awkward; simple metal tape was used instead.

This photograph details the rough terrain of Northern Ontario that still exists in the area today. The survey crew was able to complete the work in the area in the space of four summers. Niven's crews were famous for cutting through more line than any other party in the province.

With an accurate survey of the area now complete, the province of Ontario sent out its geologists to explore and map the region. E.M. Burwash and W.A. Parks found gold showings in veins along the Algoma-Nipissing boundary line, and other indications of base metals in the region, but the real work of prospecting would have to wait a little longer.

The province decided that in order to assist in the development of the northern territories, a railway line would be invaluable. While construction of the railway was underway, a silver discovery was made at a place on the line that would become known as Cobalt. The rich find encouraged prospectors fresh from the Klondike to try their luck in the north. This photograph shows three prospectors in 1905 at the Cobalt Camp.

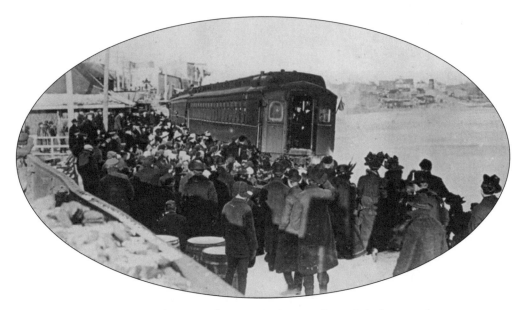

In 1906, a mere few months from the initial silver strikes, Cobalt was a busy community with full rail service and an influx of fortune hunters. This photograph shows the train arriving from Toronto with no shortage of passengers and greeters.

By 1907, Cobalt was a bustling community, with a population that was to top 20,000 residents. It became a jump-off point for prospectors who were moving northward in search of more silver or possibly gold. The railway was also moving north with the hopes of reaching Cochrane and eventually James Bay.

Two
The Rush is On!

The big rush started almost immediately after the news that gold had been found in the Kirkland Lake area. The first recorded claims in the Porcupine were made out to A.L. Phillips in July of 1908. He was quickly followed by a number of ex-Klondike veterans who staked areas around Night Hawk (over 100 claims were recorded but nothing ever developed along the river). Others moved around to Porcupine Lake and still others to the pits sunk by D'Aigle and his party. But how did these people get there? The gold finds were not developed in 1901 because of the difficulty in getting through the bush—what had changed in 1908-1909?

After Burwash's initial reports, the lure of possible mineral deposits, and therefore possible wealth, was enough to push the province towards the development of a railway. The Temiskaming and Northern Ontario Railway was created in 1902 to run between North Bay and New Liskeard, as an attempt to support the fledgling agricultural community along Lake Temiskaming and to allow the Railway Commission an opportunity to control the exploration and settlement of northern Ontario. Silver strikes at what is now Cobalt encouraged the rapid completion of the railway and started a mineral boom in the area. Eventually, the old Klondike prospectors sniffed out the stories about possible gold in the area and started the move up towards Kirkland Lake and Gowganda. The railway kept up with the men when staggering discoveries were made in those areas.

Prospectors who came pouring into the Porcupine Goldfields took the train to Kelso and then relied on a series of waterways and portages to reach the Porcupine. Hill's Landing, near Hoyle on the Porcupine River, served as a jump-off point for the trip. Many freighting companies sprang up to handle the traffic between Kelso and Porcupine, advertising their rates and "degrees of comfort" in the *Cobalt Nugget*. Enterprising businesswomen who also provided companions for the three-hour trip to the Porcupine operated a number of the stagecoaches on the route.

There is no shortage of stories about the prospectors who made their way into the Porcupine. Life was not easy; they carried 75-pound packs through muskeg and uneven ground. Of course nothing was certain—who knew what it was really like out there and what anyone could seriously find before the muck and rock was frozen over and buried in the snow? The inevitable partners—grub stakers who made small fortunes off other prospectors (but rarely ventured in the bush themselves)—were always ready to help you out for a nominal fee and a share of the take.

Real prospectors put up with all of this, plus black fly season (not to be scorned if you've never ventured out into the bush in June), bears, and wolves that were always on the lookout for an easy meal, unpredictable rivers and streams, weather that could change in a heartbeat and rancid, unappetizing food. The lure of gold, a cliché today, was, and remains, a seductive force for those who were obsessed with finding "the mother lode."

An impressive list of men and women who flocked to the Porcupine could be drawn up for this chapter; suffice it to say that three prospecting parties continue to command the respect of people in the field even today: the Wilson group who stumbled upon the Dome, the Hollinger team who were to discover one of the world's biggest gold mines, and the Sandy McIntyre-Hans Buttner partnership that found the McIntyre Mine.

Jack Wilson, grubstaked by W.S. Edwards and Dr. T.N. Jamieson of Chicago, led a band of prospectors who spurned the finds at Night Hawk and moved west. The party was not very successful in its initial forays; they pushed on to Tisdale Township however, staked a very rich piece of ground, and discovered the Ida Maud vein, which would become the core of the Dome Mine. While Wilson returned to confer with Edwards in Chicago (and probably secure enough cash to pursue the expedition), his men discovered what would become known as the "Golden Staircase," just beside the Ida Maud. The find was so rich that the gold literally stood out on the rock like huge sponges as large as cups. Harry Preston hurriedly staked the ground (even though all of their licenses were expired) and guarded the area with his shotgun. Not a man to be shy with a rifle, everyone accepted the law as laid out by Preston.

When Wilson returned, he brought Edwards with him; Edwards, unused to the rigors of life in the bush, upset the canoe and twisted his ankle. Luckily, his pockets were deep and all was forgotten when they glimpsed the find. Wilson had brought in new licenses and since the first set was illegal, a new partnership was formed. Edwards and Jamieson received fifty percent interest, Wilson ten percent and the prospectors would receive forty percent. Had Preston and his crew held the proper licenses, the story of the Dome would have been very different!

Benny Hollinger and Alex Gilles had passed the Wilson crew a few days after the find. Intimidated by the crazy prospector with a rifle, they moved even further west and staked an area that was near the pit started by Reuben D'Aigle. They were immediately rewarded with a large showing of free gold that would become the richest gold mine in the camp. As they were representing different interests, the partners tossed a coin to determine who would stake what claims. Hollinger took the discovery claims while Gillies staked six claims to the east of the find. They staked a single claim for their absent partner, Barney McEnany, which would prove to be a very generous gift.

Noah Timmins, a speculator who made a fortune at Cobalt, was elated to hear the news coming out of the Porcupine. He boarded a train in Montreal and headed to Haileybury where he painstakingly negotiated an option on the Hollinger find. He eventually developed the Hollinger Mine and started the small mine site that would become the city that bears his name.

However, back in October of 1909, the great discoveries were still underway. Sandy McIntyre had found his way to the Porcupine after deciding his life in

Scotland (as a married man and iron worker) was not as appealing as originally proposed. He hooked up with Hans Buttner and together they made their way into the Porcupine, passing the celebrating Wilson party and investigating the discovery made by Hollinger and Gillies. Duly impressed with the Hollinger find, McIntyre and Buttner hurriedly staked anything they could around Benny's claims. While the gold at the McIntyre ran deeper than that of the Dome or Hollinger sites and was less accessible, the mine would eventually become the third largest producer in the camp.

While these men all made phenomenal discoveries in the Porcupine Camp that would drastically alter the history of mining in Canada, they did not all gain fortunes from their finds. Jack Wilson and Alex Gillies lost most of their money on wheat speculation; Benny Hollinger died at the age of 34, from heart disease. Harry Preston and most of the prospectors from the Dome went broke. Sandy McIntyre did not capitalize on his find (trading his shares for about $300), and continued to live the prospector's life in northern Ontario.

Exploration up the line was encouraging after the discovery of gold at Kirkland Lake. Mining operations sprang up quickly around the area, pushed on by men like Harry Oakes and Fred Larose. The Wright Hargraves Mine was the second producing mine in Kirkland Lake; the first was the Tough-Oakes-Burnside.

In the meantime, Cobalt was still prospering. In this photograph, crowds have gathered to look at a rather well hidden 1700 pound nugget of pure silver pulled from the Nugget Mine. Finds like this one only encouraged the true mine finders to continue with their search.

"Show and tell" was always a popular event at Cobalt. The silver bricks, ready for the mint were put on display for curious residents and any potential investors to see. The silver was shipped to the Canadian Mint by rail, with very tight security.

Barges were used to ferry the ore across the lake in Cobalt. Horses helped the barges along; the load was destined for the mill and refinery. This photograph was taken in 1907; mining was already beginning to take its toll on the landscape.

In 1909, Gowganda, near Kirkland Lake, was the place to be looking for silver and gold. Fred Thomson, along with Bob Gamble and Sarkis Markarian, started the first camp. Again people moved north from this mining camp following rumours of gold.

Haileybury, nestled between Cobalt and New Liskeard, was the administrative center for the district. The all-important mine recorder's office was located there, as well as the courthouse and jail. There were many thriving businesses in the area; fine goods, restaurants, and "cultural pursuits" were all available in the town.

Large residences and summer homes were built along Lake Temiskaming in Haileybury. Many people who made their money in Cobalt had summer homes here; most of these homes still line Lakeshore Drive.

All of this exploration reassured the powers at the Temiskaming and Northern Ontario railway that lines should continue northward. Railway camps were always busy centers; in this photograph, the camp cooks are at the back, the engineers (wearing ties and jackets) and labourers to the front. The wooden buildings were bunkhouses, a kitchen, and office.

Of course, before the railway, the rough roads were the only way to get through. In this photograph, the road north from Gowganda is snow covered. It was probably easier to negotiate in the winter, when sleds could glide along easily. Carts slogging through the muck of spring and fall weather took longer.

Many temporary camps were set up along the way to house travellers to the camp. These little camps were oases on the rough road into the Porcupine. They also allowed men to exchange the gossip of the day, strike new deals, and pick up new partners.

Hill's Landing (known as Hoyle today), a stop on the Porcupine River, was a popular half way meeting point for prospectors. Here they could obtain food, a place to sleep, and stables where their horses could be watered and put up for the night. Supplies coming in by boat were transferred to wagons at this spot and hauled into the Porcupine Camp.

Horse teams hauled provisions into the Porcupine district during the winter and summer months.

When horses were not used, the preferred method of transportation was canoe and portage. In this photograph, three prospectors, McGuiness, Cullen, and Whitewell prepare to rejoin the Porcupine Trail. They are travelling with two canoes and packs for each person.

Travelling in the spring could be a "wet experience". Here, Whitewell is setting up camp; drying out packs of clothing and blankets could be a losing proposition. The fire would also be used to cook whatever delicacy was on the menu for the evening: beans, hard tack and if they were lucky, fish or wild game.

This prospector is preparing to board the train into Kelso. With him are his provisions for the summer months: tools, tarps, blankets, cooking utensils, and some food. His pack could weigh well over 75 pounds, and had to be carried everywhere.

Groups of experienced, and inexperienced prospectors came flooding into the Porcupine camp in early 1909, encouraged by reports that gold had been discovered on Night Hawk Lake. With better transportation systems in place, the way in and out would be much easier. These men are having a bite before getting back on the trail.

A traffic jam on Night Hawk Lake showed that the beginning of the rush into the Porcupine was on!

Portaging was a rigour of the trail that no one escaped. Supplies had to be carried overland when the creeks and rivers became impassible. This could necessitate many trips by foot back and forth along the shoreline. This gentleman chose to bring along a second pair of hob-nailed boots—essential for the rough terrain!

With important discoveries being made all over the Porcupine Goldfields, it was unthinkable to give up the hunt just because winter set in. This group of prospectors is readying their sleds, braving a wicked wind.

Some prospectors set up a permanent camp in the goldfields. They built log cabins and settled in lodgings that were palatial, compared to the rough tents and shacks along the Porcupine Trail. Prospectors supplemented their income with a little trapping; notice the pelts hung along the window.

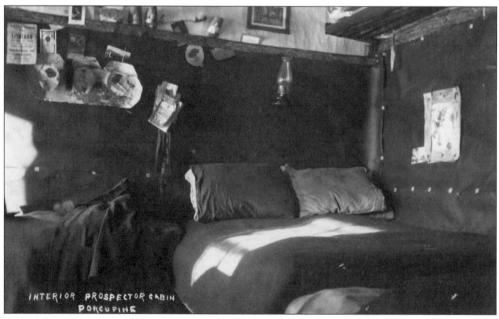

All of the comforts of home—a soft bed, light, photographs of the girls back home, and insulated walls. "Bill" sent this photograph home to his mother who lived in Rochester New York; no doubt she was worried about her son up in the wilds of Canada.

When horses were not available, men coming into the camp used dog teams. This team could pull the canoe along the frozen lake, in anticipation of the spring thaw. Dog team racing became a popular sport when the camp was settled.

One of the successful "fringe businesses" of the rush was freighting. Getting supplies into the area was a challenge and many people specialized in opening up the area for trade. The Butch Burns Freighting Outfit did a brisk business on Porcupine Lake.

George Bannerman staked the Scottish-Ontario Mine, several miles north of Porcupine Lake. Many of the men going into the Porcupine passed by the mine and had a look. That was all they needed, to know they were headed for gold country. Although the showings were impressive, the Scottish-Ontario did not live up to its billing when the project went underground.

Men were not the only ones flocking to the camp to find their fortunes; women took to the trails and also prospected. They also came to the camps as cooks and entrepreneurs. This woman prepares for the trail in her beaver coat and snowshoes.

Caroline Mayben Flowers, of New York City was a fixture in the Porcupine Camp. During the week she prospected in the area surrounding Porcupine Lake, but never actually recorded any claims with the Mine Recorder's Office. On Sunday mornings, she played the organ for the Methodist services in the makeshift church at the pool hall.

Johnny Jones was a well-known character in the Porcupine Camp. He once sent a dog sled loaded with Hudson's Bay coal down Yonge Street in Toronto as a present for the mayor. He kept over 50 huskies in his kennels at Coniaurum Mine and raced the dogs across the country. Johnny never found any gold in the Porcupine, but had a great time trying.

This group was one of the first into the Porcupine at the beginning of the rush. It includes George Bannerman who staked the Scottish-Ontario mine, and Whitewell, who joined them along the trail. Others pictured include Misters Bele, Hughes, and Webb.

Bannerman's Camp became a fixture in the early days of the rush. Their success encouraged others to keep up the hunt for gold. Unlike many prospectors who ventured into the Porcupine, Mr. Bannerman stayed here and made numerous contributions to the community; he served as Whitney Fire Chief and as Reeve of the township. His partner Tom Geddes died during the Porcupine fire while trying to save his dog.

Jack Wilson led his party, financed by W.S. Edwards of Chicago, into the camp in 1909. They set up this office in the winter of 1909 on the present-day site of Dome Mines. His group included Jack Miller, Tom Middleton, Clary Dixon, and Harry Preston. Jack Miller decided not to accompany the group into the Porcupine, but provided them with a canoe and supplies for the trip that Clary Dixon believed did not add up to $100. Miller was promised a part of whatever they found.

The camp they initially set up was a little rough. The find they made however, would change the future of that piece of land. Festivities aside, the boys were a bit leery about their claims, as their licenses had run out and they had to make everyone believe the stakes were proper. The matter was cleared up in some fashion when Wilson acquired new licenses in Haileybury.

Pictured from left to right are, Clary Dixon (stepson to Jack Miller), Jack Miller (who made a fortune without much effort), Alex Gillies (who would later join Benny Hollinger), and Tom Middleton; these men discovered the Dome Mine, which is still operating in South Porcupine.

Alex Gillies and Benny Hollinger were the co-stakers of the Hollinger Mine. After Hollinger made the discovery in October 1909, he and Gillies immediately staked twelve claims for themselves and one for their friend Barney McEnaney. His single claim became the nucleus of the Porcupine Crown Mine. Legend has it that Hollinger and Gillies flipped a coin to divide their claims. Hollinger won those six claims on the west side of the find.

In 1910, the first camp on the Hollinger claims was set up. Pictured from left to right are, Mr. Reid (a prospector after whom a lake is named in the area), Jim Labine (a business partner of Henry Timmins from Kirkland Lake), the Cree guide who led them to the camp, Alex Gillies (staked the claims), Noah Timmins (who bought the Hollinger claims), and R.G. Campbell (a member of the Timmins party).

This portrait of Benny Hollinger at the age of 34, was taken in 1919. He died shortly afterwards from heart disease at his home in Mattawa.

Sandy is perhaps the most colourful of all the adventurers who made it to the Porcupine. According to one story, after he discovered the McIntyre Mine, he left the bulk of his newfound wealth with a local banker. Sandy promptly returned to Scotland where he tried to drink more Scotch than his native land produced. After a bout in the army, he returned to the Porcupine to find that his stake had been lost in speculation. Undaunted, he returned to his first love—prospecting.

Sandy's partner in the McIntyre claim was Hans Buttner, a German prospector who made a small fortune on the mine. He is pictured here (back row, left) with his family in Germany.

A windlass was used when the first pits for the new mines were dug. It allowed rock from the bottom of the mine to be raised to the surface. The sides of the pit were lined with wood, and eventually tunnels could be cut out from the original pit.

All of the camps eventually had cooks working with the men. These two gentlemen were probably with the Bannerman camp.

GLORY HOLE DOME MINES, SOUTH PORCUPINE. ONT.

The Glory Hole was the beginning of the exploration cycle at Dome Mine. The area is still being mined today. This hole was the result of blasting the Ida Maud and starting the work on the dome of rock.

VEIN OF GOLD 4 FT. WIDE ON BIG DOME

Father Paradis and an unidentified member of the Wilson party are sitting on the Big Dome. At that point in the rock, the vein of gold named the Ida Maud for Jack Wilson's wife, was 4 feet wide. The Dome was eventually blasted through to create the Glory Hole.

Father Paradis and the Wilson crew examine the Golden Staircase at the Dome Mine. It was situated beside the Ida Maud, and was described as "a terrific show, gold stuck in the quartz all over the place, like candle drippings."

Once the discovery was made, and gold was seen on the surface, the vein itself was stripped and prepared for mining. This group is working on a vein at the McIntyre Mine. At that time, the tools of the trade were shovels and picks.

The finds in the Porcupine Camp were huge. There was barely any discussion about extending the railway and creating a spur into the camp; gold had been discovered in Ontario and contrary to popular belief it ran deep. Many more people would flood into the goldfields in the next six months and transform the rough camps into small towns.

No one can deny the lure of gold and the power of promised fortunes! These men are seen working on the Mattagami Wagon Road in the Porcupine Camp. The spring melt, coupled with muskeg and dense bush, made the work hard and seemingly impossible, but the anticipated reward kept everyone on the job. As pictured, most of the heavy work was made easier by draft horses.

Three

The Porcupine Goldfields

While the early prospectors may not have made their fortunes, they had cleared the way for the development of the mines. In early 1910, a rough road was cut from Kelso to the Porcupine. This new road brought in experienced mining men and drifters looking to make a few dollars. Kelso grew into a bustling community almost overnight—tarpaper shacks, stables, transport businesses and the like became home to hundreds of people in a few weeks. Lillian Beatrice Knapp, a pioneer of the Porcupine, recalls her trek into the camp in 1910 on the heels of Noah Timmins:

We left Haileybury on January 9, took the train from Haileybury to the end of the steel. We had our men, also three teams of provisions. We stayed at a hotel in Kelso. I believe its name was the Cochrane Hotel. That night every room in the place was taken. It looked as though I would have to spend the night sitting up until a young man came up and gave me his room. I have often wondered where he slept that night.

Here I was the only woman in our outfit. We left at the break of day and started on our long drive. It was awful cold and snowing some and riding on top of the provisions did not help any. When I got too cold I would start walking until tired, then back up on top of the provisions.

On the trail there were a few tents selling hot coffee. I guess we made a stop at all of them. By that time, we were pretty tired, also the horses had a tough time as our loads were pretty heavy and at some places the trail was not wide and the men would have to cut down a few trees to go on.

We got to Father Paradis' [Connaught] to feed the horses, also to rest, and by that time we were all so cold and hungry. Father Paradis made us a good dinner and had me have a couple of hours of rest, but like everyone, we wanted to get to the end of our journey. The next day we crossed Porcupine Lake and put up our tents and cook tent. The next day more people were coming with dog teams, also a lot of them walking. Everyone was out to stake their claims. [Frank Lendrum, "Days Gone By," *Timmins Daily Press*, February 5, 1949]

Winter transport was easy compared to the spring quagmire that awaited the pioneers. The road from Kelso to the Porcupine was about 24 miles long in the winter, and 52 miles long in the spring, because of the road and water route that had to be used during the breakup. It is easy to glide on snow and frozen lakes, but a nightmare to have to trudge through the mud in the bush. Tree stumps did not make it easy for the horses that had to pull wagons over the rough terrain. But gold was discovered and it looked like a rich haul, so the provincial government quickly used prison labour from southern Ontario to build the roads into the area. Talks also started about extending a branch of the railway to the new mining camp.

The summer of 1910 saw the creation of seven town sites that grew around the original outposts of Golden City (on Porcupine Lake) and Shuniah (Pottsville). While rough houses and businesses (and the ever popular saloons) sprang up in South Porcupine (across the lake and closer to the developing Dome and Hollinger mines), Golden City remained an important stop for prospectors because it was home to the mine recorder's office. Shuniah held on until 1911 because it had the area's first post office in the Réveillons Frères store.

The new mines required lumber for the underground workings and power to operate the equipment. The managers of the Hollinger were quick to realize that the mines should provide these facilities. A sawmill was built on the Mattagami River, while a power station was developed at Sandy Falls, a few miles from the mines. Hundreds of tons of material, and hundreds of people were moved into the camp in 1910. Development was rapid in the little towns. Businesses grew, houses were built, families started to move in and a feeling that maybe the camp would last longer than a few months encouraged the residents to lay down roots. Events in the summer of 1911, however, would hinder this new philosophy and challenge their will to remain in this isolated outpost.

Business prospered in Kelso at the beginning of 1910. The many people who ventured into the area needed a place to stay; the train stopped at Kelso and stagecoach operators took over and drove them to the Porcupine, if they weren't already set up with horses, dogs, canoes, or sleds of their own. The boarding house at Kelso offered clean sheets and warm food.

All manner of supplies were brought into the camp by horse-drawn sled in the winter and carts in the summer. Freighter canoes and barges made water crossings easier. These horses are bringing in lumber for the new camps.

In the winter, stagecoaches were exchanged for sleds that allowed their passengers to travel in style. They probably would have been more comfortable if the sled was covered! The passengers are wearing beaver coats that would have helped to keep them warm.

Jack Dalton came to the camp in 1910. He started the first livery service the next year; he provided a freight service and a "taxi service" between the camps. He is seen in this picture at the reins, bringing in the Timmins party.

The town of Golden City sprang up early in 1910 along the shores of Porcupine Lake. With the influx of people, all sorts of amenities were needed. The Gold Mint Saloon and Pool Room did a brisk business (attested to by the amount of empty kegs seen in the snowbank); the saloon was used as a church on Sundays.

This pool player is enjoying a game in the Gold Mint Saloon. The crucifix on the wall must have been left over from the Sunday service.

Other businesses soon grew up in the camp. The tent seen in this picture was the temporary post office. The streets were still filled with stumps and in some spots, full-grown trees.

Wood-framed buildings also made their appearance in 1910. The Porcupine Office of the Wee-Tu Mining Company, Welch and Son Real Estate (the owner of the building), the Cobalt General Store and J.P. Crawford Law Offices all shared space.

Golden City was established as the administrative centre of the camp. It already housed the post office (official recognition of the existence of the camp by the federal government); now it became home to the all-important Mine Recorder's Office. Prospectors no longer had to run to Haileybury to record their claims; it was conveniently done across the lake.

The Shuniah Hotel in Porcupine was the last word in luxury in the camp. It provided accommodations for the speculators who came to check on their investments. The hotel was a large two-storey affair, built out of logs, owned by Ma and Pa Potts. In the background, is the log cabin office of the Porcupine Telephone Company.

The Shuniah Hotel also provided transportation for its guests. In this case, seven dogs are pulling a very full sled.

By the late fall of 1910, Porcupine City was well established. A hardware store, a general store, and small homes added their bulk to the street scene. In front of the hardware store, a street lamp has been erected to keep things bright in the early dark evenings.

The southwest end of Porcupine Lake was also under development in 1910. If they could maneuver through the street, travellers could enjoy a lovely lunch at the King Edward Dining Room (housed in a tent) and stay in the "first class sleeping camp." They could also enjoy a game of pool and a good cigar at the Palace Pool Parlour.

Across from Golden City, a rival community was taking shape. South Porcupine was becoming a favourite spot for those who wanted to be close to the developments at the Dome Mine. The tent and log cabin community soon would overtake Golden City as the center of the Porcupine Camp.

Real estate was suddenly in great demand around the camp. Real estate agents and brokers began to set up shop in Golden City and Porcupine. The Real Porcupine Co. opened in 1910 and continued to do business until the fire of 1911.

Another important amenity to be had in the Porcupine camp was an assay office. The Porcupine Assay Co., owned and operated by Hatch, Duncan, and Hyland saved prospectors the long trek to Haileybury to have their finds assayed and valued.

GOLDEN AVE. BEFORE THE FIRE. SOUTH PORCUPINE

In South Porcupine, the Christakos Brothers set up shop and moved their family to the camp. The shop was a general store/outfitter operation that was located on what would become Golden Avenue.

SOUTH PORCUPINE
BURNED, JULY, 11, 1911
TOMKINSON PHOTO

It didn't take very long for the tents and log cabins of South Porcupine to evolve into a small town complete with framed buildings and competing merchants. While the streets still needed work (stumps can still be seen in the centre of the road), boardwalks near the buildings made walking easier.

Down the line a bit further near the McIntyre Mine site, the small community of Aura Lake was settling down. F.W. Schumacher was instrumental in setting up the community; the town would soon change its name to that of its founder. By the next year, it would have a train station of its own and a town site that would rival those around Porcupine Lake.

Four

The Great Porcupine Fire

During the spring of 1911, the bush and trails around the Porcupine Camp were not the usual black stuff; the muskeg had dried into a brown, tinder-like material that ignited regularly. Cries of "Fire!" brought people running into the streets with every imaginable implement to douse the flames. Caught in the middle of a powder keg with only one slow way out (for the railway was not yet completed), the miners learned to be vigilent. In May of that year a fire swept through the Hollinger mine site, destroying every building, head frame and piece of equipment. It is only because of Noah Timmins' determination that the mine was not abandoned and work was started almost immediately on re-building.

As summer approached, rain remained scarce; the bush was completely dry. Fires continued to threaten the towns and mines. The 200 employees of the Dome were able to save their mine in June, but nervous residents were quick to realize that the fire had been less than one mile away. They also realized, however, that this was an accepted way of life in the north.

While this drama unfolded, the railway was nearing completion. The first passenger railcar arrived in the camp on July 1st, 1911; the gold mining communities were officially on the map. Life would be a lot easier from now on, or so they reasonably thought.

That first week of July in 1911 brought home the extreme nature of northern weather. Winter had been bitterly cold, and now a summer heat wave threatened to cook the rest of the forest. A grey-blue haze covered the sky on July 10th; everyone knew that the bush was filled with small fires in the muskeg, but a refreshing breeze was picking up, and people were confident that it would start raining soon.

That gentle breeze however soon developed into a gale force wind, encouraging the smoldering fires to explode into the bush. The early morning of July 11th saw a mile-long band of flames headed towards the new works at the Hollinger Mine; the workers recognized the danger and headed towards the newly arrived railhead. In the towns around Porcupine Lake, people quickly realized that this was not a small fire that could easily be put out; they gathered some of their belongings and hurried towards Porcupine Lake. A jeweler buried his wares in a trunk in his backyard; barges at the docks were loaded with women and children. Men ran into the lake with bears and moose who were also running for their lives. Countless stories from the people in the area were recorded. John Campsall was a young boy newly arrived in the camp, and years later had this story to tell:

> Now I will tell you about the fire as I remembered it. It was the day before my eighth birthday and I was standing on Queen Street with my mother and sister, watching the houses catch on fire one at a time as the fire came up the street, and wondering when our house would burn. I am not prepared to say how far down the street I saw the houses burning, but it would have be as far as Third Avenue anyway. I have read a lot of different versions as to how much of Golden City was saved…We were waiting beside our trunk while my uncle went to get a team to haul it to the lake. The wind was blowing with almost hurricane force and the air was literally filled with smoke, sparks and ash. Just about that time the wind tore the lid off our trunk and sent it sailing up the road towards Porcupine station with some of our belongings. Mother took us by the hands and started towards King Street. Just as we reached King Street my shirt caught fire. Mother managed to put it out and started to cry. One of the few times I ever saw mother cry. Just about then a man came up to her and said "Lady, you better head for the lake"…When we got there most of the townspeople were already there. Talk about confusion! Many people were trying to cross the lake from South Porcupine to Golden City, as South Porcupine was a solid mass of flames. Some of them were drowned due to the high wind.
>
> After the fire we went back up the street to see what we could salvage. Some blankets which we had put in the well had holes burned in them when the fire followed the wooden cribbing down the well. A bag of flour which we had put in a ditch and covered with clay was all right. A roast of meat which we had put in the oven was nicely done when the house burned down. [John Campsall, *Tales of the Early Porcupine.*]

The official death toll is recorded at 77 people; accounts do not include any prospectors or bushmen who may have been caught by the flames outside of the towns. Robert Weiss, manager of the West Dome Mines, took his wife, daughter, and 22 workers into one of the shafts to escape the flames. They died of suffocation when the flames raced over the hole and drew out all of the oxygen. Many others died when a railcar filled with dynamite exploded; the debris rained on those standing in the lake or along the shore, and the shock waves churned up the lake, drowning many.

There are countless tales from survivors of the blaze; a group of miners sat in a small creek and were saved (it became known as Salvation Creek). A woman gave birth to her daughter in the lake. John Munroe led a bucket line in a futile attempt to fend off the flames.

The Porcupine was wiped out; eleven mines were completely destroyed. Not one business or home was left standing. The telegraph service was gone; the new railway that just days before had been a source of pride was now a series of twisted metal beams. Luckily, on the next day, a train was able to make it in far enough to bring in relief supplies. It also took out the orphans, widows and some of the dead. It would take only a few days for the Porcupine to start along the road to recovery. The gold finds were just too important (and too big) to abandon.

Work on the spur line into the Porcupine was hastened in early 1911. The mines were ready to start production; materials and supplies were needed to assure the success of the camps. The population had grown to 3,000 and a rail line would assure the future expansion of the area.

The T&NO crews met their deadline and the first passenger train to Golden City arrived on July 1st, 1911. Many people gathered at the station to welcome the official delegation. The future of the new towns seemed assured.

This famous photograph by Harry Peters appeared in the *Globe* newspaper in 1911. The only problem is that it was faked. Peters was unable to get a dramatic shot of the town burning; he created this composite by superimposing two photographs. We know it is a fake because the telephone poles have been drawn in, the people don't seem to be very panicked (considering they are standing at the end of the street nearest to the blaze), and the horses in the foreground seem rather calm. The photograph did however impart the terror of the blaze and helped to organize the relief effort out of Toronto.

A bucket brigade was set up early in the morning and continued to run until mid-afternoon. Jack Munroe organized the men and fire wagons, but they had to give up when the temperature reached 118°F. While they put in a gallant effort, the buildings of South Porcupine were all lost.

All available horses were used to pull fire wagons and regular wagons equipped with barrels. Horses made the trip to the lake where the wagons were filled with water, they were then led back to the town and the fires. Eventually, the horses stayed in the lake to escape the flames.

DURING PORCUPINE FIRE AT GOLDEN CITY.

The last train had left the Porcupine hours before the fire started. There was no way out of the camp, so frightened people headed for the lake. Barges took women and children to Golden City. Luckily for others, the lake was not deep near the dock and people could wade in for about one hundred yards and keep cool in the water.

The fire burned through the camp quickly, taking everything in its path. The next day, a number of homeless refugees were brought out of South Porcupine. Orphaned children, widowed women, and others who chose to leave were moved out by rail.

Survivors started to salvage items from the camp and count their blessings. Tents and temporary shelter went up immediately as people tried to gain some understanding of their plight.

Thanks to the Peters photographs and the quick relaying of information, relief trains from Toronto poured into the camp. They brought much needed supplies to the camps and transportation for those who wished to leave. They also brought out the bodies of those who perished.

This eerie sight greeted the survivors of the fires in South Porcupine. All of the buildings and streets were destroyed; the telephone and electrical poles were the only things left standing.

Robert Weiss, mine manager for the West Dome Mine, died with his family and workers in the mine's main shaft. They are buried out at Dead Man's Point on Porcupine Lake. One man, Rosaire Bourbeau, the mine's accountant, did not join the group and survived.

The West Dome Mine was already in production when the fire came through the camp. No buildings were left standing, and the shaft leading into the mine was exposed. Twenty-five people lost their lives here when the fire passed over the shaft and drew out the oxygen from the hole.

Lumber to build coffins was brought into the camp on the train. Work parties had to make 77 coffins for the known dead; the death toll may have been as high as 250 because no one really knew how many prospectors, lumbermen and others were in the bush that day.

The dead who were to be buried in the camp were taken to Dead Man's Point across from South Porcupine. This quiet and serene point of land juts out into Porcupine Lake. A monument erected by the Toronto Board of Trade commemorates those who died.

To compound the problems faced by the residents of the Porcupine, a rail car containing dynamite for the mines, newly arrived in the camp, exploded from the extreme heat. The blast tore up the new track and created a huge hole in the ground. The force of the explosion was felt throughout the camp; boats on the lake capsized and many people drowned.

Many stories of survival came out after the fire. This group of men submerged themselves in this small creek and lived to tell their tale. The small waterhole became known as Salvation Creek, as it saved 20 lives on that day.

These men were saved when they found themselves in Porcupine Lake. Their canoe was overturned because of the blast from the exploding dynamite car, they were able hold on to their canoe and drift in to Golden City. Billy Moore, the owner of the canoe, did not survive the ordeal.

The fire around the camp had cleared the bush. This prospector surveys the area around his home that was also destroyed.

The gold mines did not escape the flames. The Dome was completely destroyed; the Hollinger had already gone through a blaze in May, and had only started to rebuild. The McIntyre Mine was also burned out. These men are posing with the remains of the Dome power plant.

The T&NO work crews were still in the bush working on the railway when the fire struck. Camp No.6 lost some of its buildings; the dining room was moved out of doors.

This reporter created a makeshift office from remnants he found. His typewriter miraculously survived the blaze. Word of the fire spread quickly thanks to the work of mining journalists, telegraph operators, and T&NO personnel.

Makeshift shelters sprang up in the camp. The temperature was still high, the air was filled with smoke and ash, and food was scarce. Even with these hardships, the rebuilding effort started almost immediately.

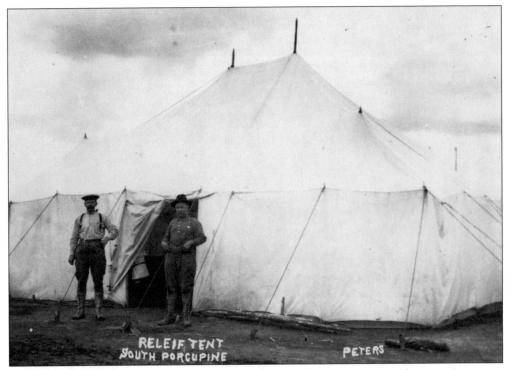

Tents were set up around the camp to accommodate the survivors. They were sent by the T.E. Eaton Company in Toronto, along with other necessary supplies.

Many refugees chose to set up their tents near the water's edge, in case the fire started up again. In the background, Pottsville can be seen; that community was spared from the flames along with Golden City.

The Canadian Bank of Commerce opened almost immediately for business after the fire. The gold mines around the community were simply too rich to be abandoned.

Jack Dalton lost his livery business in the fire, but was able to rescue his horses. Both of these animals were saved because they stood in the lake; they did suffer burns to their backs. Dalton's business was up and running in a few weeks after the fire.

The temporary post office in South Porcupine was delivering mail within days of the fire. A return to normalcy was the goal of the residents, every little bit helped.

People buried many of their valuables when they saw the flames approaching the town. This young man is retrieving his valuables that were buried in a trunk; Mrs. Campsall found her bag of flour that she had buried in her backyard.

The clean-up in the camp would take the rest of the summer months. By the fall, the residents had new homes and shopkeepers new shops. By early 1912, the destruction caused by the fire was a memory for most of the people in the camps. They were prepared to move on with the development of the gold mines and the rebuilding of the communities.

Five

All That Glitters

One good thing to come out of the fire was that the bush around the Porcupine was stripped down to the rock, allowing prospectors a clear view of the land. Many flocked back to the area and began the discovery process all over again. The Dome, Hollinger, and McIntyre mines were finding large ore bodies around their sites, and they rebuilt their plants and facilities.

Although many people chose to leave the camp and the harsh life that it entailed, many stayed to live in the new communities. Tents and temporary shelters replaced the clapboard buildings during that winter. Banks conducted their business under canvas; so did many merchants. The T. Eaton Company sent supplies to the camp and the help from Toronto was instrumental in restoring a sense of confidence in the pioneers.

The Hollinger Mine was recovering, even though it had gone through two serious fires and a loss of equipment that was staggering. Noah Timmins knew that he had to create another town site for his expanding workforce, somewhere nearer to the mine. In the autumn of 1911, barely eight weeks after the Porcupine Fire, an auction was held near the temporary offices of the Hollinger mine. Lots were being sold for prices between $5 and $1500; buyers with active imaginations could imagine the new business district in the empty land before them. Others found the prices too high for a thirty-foot chunk of muskeg. Timmins, ever the dreamer, enticed buyers with plans for a luxurious hotel, homes for employees, and bustling shops. The railway station would be at the top of Third Avenue, moments away from the mine and the businesses.

In no time buildings began to appear and businesses moved in. A petition signed by 78 landowners requested that the "town of Timmins" be incorporated; it was approved in December of 1911, and took effect on January 1st, 1912. The community grew rapidly with the aid of the Hollinger Mine; in the spring of 1912, there were over 500 people who claimed residency in the town. The Goldfields Hotel was built near the train station; substantial homes were built for the mine management on a hill near the town; many of those homes still survive today. That year, just three years after Benny Hollinger and Alex Gillies made their discovery, the Hollinger Mine paid dividends to its shareholders. The Dome was performing on the same level and the communities looked solid.

Meanwhile, F.W. Schumacher purchased eight acres of land between the Hollinger and the Dome properties, near Pearl Lake and the McIntyre Mine. He had the land surveyed into building lots and founded the town which stills bears his name. The community grew quickly and came to rival South Porcupine in size. Boards of Trade in Timmins and South Porcupine developed campaigns to lure investors and miners to their communities. The South Porcupine Board of Trade printed a list of impressive accomplishments and reasons to visit the camps in March of 1912:

> Only 36 hours from New York and Boston. Less than 20 hours from Toronto.
> Scores of working mines. Thousands of miners on the payroll.
> Barely 2 years old. Over 10,000 people.
> Great mills, great power plants, and other buildings of brick, steel, and concrete construction.
> Electric light and power furnished by two great water powers.
> Telephone system connecting six town sites and all the leading mines of the district.
> Railway transportation right into the center of the camp. Two Pullmans daily.
> South Porcupine, the business and financial center for the district is destined to become the Metropolis of the North. Electric-lighted and steam-heated hotels; business blocks; six banks; wide streets; water and sewer systems; progressive, hustling, up-to-the-minute business, professional and mining men are here and we have just begun to grow!
> Come up and have a look. Business is good.
> Gold is the basis of all the world's wealth and we have the gold. [South Porcupine Board of Trade, "An Occasion of International Importance – Porcupine's First Gold Production", March 1912.]

Miners did come north, and so did many businessmen and women. Confidence had returned to the community, as well as a sense of invulnerability. However, not everything goes to plan all of the time.

In November of 1912, a general strike of the miners of the Porcupine District was called. The origin of the dispute lay in the decision made by the mine managers and owners to put into force a reduced rate of wages. About 1200 men were out of work at the beginning of the strike; many could not afford to stay in the camp and left. When the mines brought in a group of strikebreakers and Thiel constables, riots broke out on the picket lines and events degenerated rapidly. Provincial police were brought into the camp to restore order after the special detectives fired on the strikers. The strike continued until the early part of 1913; the union gave in to the employer's demands and production at all of the mines resumed quickly.

By July of 1914, the towns of the Porcupine Goldfields were successful enterprises. Train service had been established as well as roads, waterworks, hospitals, theatres, hotels, shops of all kinds, and a newspaper. Vaudeville acts were making their way to the area; the mines set up a number of different sports teams. Friendly competitions on the rugby fields and baseball diamonds were part of Sunday afternoon activities. That summer, life would change dramatically for the people of the Porcupine, just as it did for all of the members of the British Empire when war was declared in August of 1914.

The men of the Porcupine Goldfields played a double role during the First World War. Miners were used extensively along the front; their underground skills proved effective when tunnels were needed. The young men who ventured north to seek their fortunes were also the first to sign up for the adventure of war. Two Canadian regiments were formed in Northern Ontario: the Algonquin and the Northern Pioneers. Patriotic Funds were set up in most communities, particularly those with strong British ties. In the Porcupine Camp, the mines continued to produce non-stop, as the credit of many countries depended on the precious metal for food, munitions and clothing. Gold mining was a favored industry during the war, and keeping the mines operating was a priority.

When it became clear that the war would not be over quickly, an aggressive recruitment campaign was launched in northern communities. The 228th and 256th battalions were formed in this area. They became part of the Algonquin Rifles at the end of the war. Over 600 men enlisted from the Porcupine Camp during the war years.

The towns continued to develop and grow during the war years. The Hollinger Mine started work on the "Hollinger House," the green and red tarpaper homes that would house the miners and their families. Churches were also built in the community. Father Theriault founded St. Anthony's Roman Catholic Church, while the Hollinger Mine assisted with the development of the Anglican Church (parishioners were beginning to complain about their temporary quarters in the pool hall). The strong Jewish community built a synagogue in downtown Timmins. Community halls were springing up in all three towns; organized along ethnic and political lines, the halls became second homes to newly arriving immigrants. These halls also served as meeting places for workers looking to organize unions; the World Federation of Miners was powerful in the camps at this time, but eventually lost support in the north to the International Union of Mine, Mill and Smelter workers.

Fires did not stop ravaging the towns. In 1916, fire threatened the outskirts of Timmins; 20 homes were destroyed, but the fire was contained. It would work its way around the Porcupine district and destroy the communities of Cochrane, Iroquois falls, Porquis Junction, Ramore, and Matheson. The scare was enough to convince the townspeople they needed a full-time fire department with proper equipment.

Life was not all tragedy and hard work, however. The *Porcupine Advance*, in July of 1915 proudly reported that no fewer than 22 automobiles were in daily use in Timmins and South Porcupine (although they could not drive to and from those communities). The roads in town were hard-packed dirt in the summer and snow in the winter, making driving a true adventure. Athletic grounds were created on Dalton Road and races were held during the weekends in the summer. In winter, dog sled races were popular and drew participants and crowds of spectators from across northern Ontario. Shops sold fine dresses and accessories for the ladies of the town; afternoon teas, dances, and performances dictated the social calendar. In 1918, the town's social butterflies were treated with a visit from the Prince of Wales. Months of preparations and the fine-tuning of the local band were rewarded with a fifteen-minute stop in Timmins and a quick ride through South Porcupine. His Majesty did receive an impressive nugget of gold from the Hollinger Mine for his trouble.

The Porcupine Goldfields were solidly established in less than ten years. The gold mines prospered and northern Ontario was no longer an unknown frontier. It was now home to newly arrived immigrants and people looking to carve out a new life in New Ontario.

After the fire, it did not take long for prospectors to regroup and start out for their claims. The fire assisted them with their work by clearing out the underbrush and revealing outcrops of ore. The men also headed back to the sites of the big three mines to begin the clean-up process.

Within a few days of the devastation, South Porcupine was back in action. Tents and temporary housing were set up for the townspeople, supplies were being sent in by rail and by barge. Those who chose to stay in the camp were prepared to rebuild their homes and make a life for themselves in this new section of Ontario.

Munn, Blacklock, and Burton re-opened their restaurant under the name "Phoenix"; like the legendary bird, South Porcupine was quickly rising from its ashes. While the concept is a cliché, this photograph became a testament to the pioneer spirit in the camp.

Barely eight weeks after the fire, Noah Timmins started to sell lots in the area around the Hollinger Mine; he hoped the new town site would grow into a thriving community. Not many people were able to share his vision, but some showed the necessary faith and purchased lots on what would become Third Avenue, the business core of the Town of Timmins.

By the time the town was incorporated in 1912, a small business district was growing up on Third Avenue. It was located near the train station, and near the Goldfields Hotel, the two main attractions in the area.

Part of Noah Timmins' dream was to create a fine hotel for visitors to the mining camps. The Goldfield Hotel was opened in 1912 and was one of the finest buildings in the area. It was a three storey framed building with well-appointed rooms and a dining room.

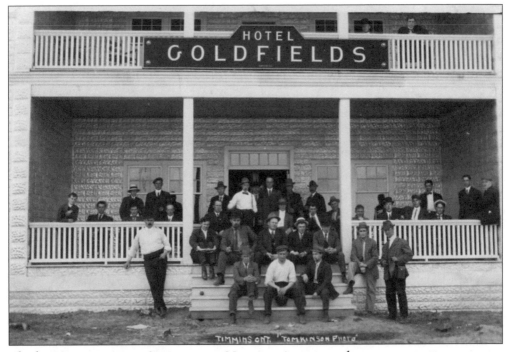

The hotel was an immediate success. New immigrants to the camp, reporters, speculators and sightseers made the Goldfields their temporary home. Noah Timmins' dream was becoming a reality.

The spur line into the Porcupine Goldfields was completed at last and a train station was built at the top of Third Avenue in the new town of Timmins. It also made travel between the mine sites much easier, from Timmins to Schumacher to South Porcupine, Golden City and beyond.

Production at the McIntyre Mine began in 1912. This photograph shows the development that occurred around the mine site; the Schumacher town site is represented by a few homes and businesses. The mine had already built a clubhouse for the miners and a bunkhouse for the single men.

This bungalow was built for the mine manager of the McIntyre Mine. On the back of the postcard, the manager's wife describes the home as "bleak outside but lovely inside."

The Dome Mine was quickly reconstructed after the fire of 1911, and was actually expanded to accommodate the increase in production. In this photograph, the mill is visible in the background; an early headframe is in the foreground.

The six claims staked by Benny Hollinger became the core for the Hollinger Mine. At the height of production, the mine was the largest operating gold mine in the world. During its operating life (1910-1968), the Hollinger produced over $566 million worth of gold.

While the big three mines, the Dome, Hollinger, and McIntyre were the stars of the Porcupine Goldfields, other properties were discovered and moved into production. The Lapalme Porcupine Mines Company developed the Three Nation Mine in Porcupine; it eventually was bought out and became the Pamour Mine that continued production until the late 1990s.

Barney McEnany was to have been a part of the Hollinger team but was ill and unable to make the journey; Hollinger and Gillies staked one claim for him and it became the core of a large mine. The McEnany went on to become part of the Porcupine Crown Mine, and eventually the Vipond.

The Dominion Store opened in the camp after the fire and was a part of the local business scene for years. H. Korman, the store's proprietor, went on to own many local businesses, to fund the building of the local synagogue, and to open Korman's Dairies.

Sam Bucovetsky and his brother opened stores in South Porcupine, Timmins, and Kapuskasing. He started out with men's clothing and soon developed into a rather posh department store. His store in Timmins is still open for business.

Bucovetsky's delivered their goods to your door. Sam Bucovetsky is second from the left on the wagon.

A number of bakeries of course set up shop in the camp. These bakers from Schumacher show off their wares for the owner.

King & Babe June 25th 1917

Jack Dalton recovered from the fire of 1911 and his livery business soon blossomed. He provided a taxi service, delivery service and general assistance to all. His motto, "We Never Sleep" was adopted as the town slogan in all their promotions.

The World Federation of Miners had a miner's union hall in the Porcupine in 1912. Most workers joined the union and fought for safer conditions in the workplace, as well as increased wages.

In November of 1912, the miners of the Porcupine called a general strike. The local mine managers were attempting to put into force a reduced wage scale. Over 1200 miners walked out and halted production at all of the mine sites.

The strikers organized a march for January 1st, 1913. It began at the Goldfields Hotel and snaked its way through the camps. The banners proclaim the well-known slogan "Workers of the World Unite" in a variety of languages.

The mine managers eventually brought in a group of private constables (Thiel Detective Agency) to provide security at the mine. Actually, it seemed the detectives were there to crack a few heads and antagonize the strikers. Tempers flared during a protest and the detectives fired into the crowd, injuring workers and causing a panic.

After the melee between the workers and the agency, the Ontario Provincial Police were called in to provide security for all involved and to restore order in the town. These fine gentlemen are pictured in front of the Goldfields Hotel.

The OPP remained in the camp and a jail was built at Porcupine. According to the Board of Trade, in its promotional pamphlet written in 1913, "the town has a comfortable jail, with modern cells, for taking care of any unruly spirits that it may appear advisable to keep in custody."

With the strike finally settled (the mine managers' scheme was adopted by the union), life returned to normal at the mine sites. This group of young men was working at the Dome Mine in 1913, either underground or on the surface.

The Porcupine Camp was not isolated from world events, and soon the young men who ventured north to find their fortunes were now looking for adventure in the battlefields of Europe. This area, by the end of the First World War had sent over 600 young men to the army. J.P. Bartleman's shop also served as the local recruiting hall.

The north contributed two battalions from the area: the 228th and 256th Fusiliers. These battalions would later form the Algonquin Regiment. The officers of the 228th are seen visiting the town of Timmins and the Goldfields Hotel.

Dome Hospital

While it became clear that the war would not be over in weeks, months or even years, life in the camp continued. The mines continued to produce much needed gold. The mine sites developed into permanent settlements that looked after their work force. The Dome opened a hospital on its town site available for workers and their families.

Since the hospital was there, staff was needed. The Dome hired a nurse and doctor to work on the site. This practice continued for years at the mine.

The Hollinger Hospital originated as a company institution owned by the mine, accommodating a total of 12 patients. Noah Timmins invited the Sisters of Providence to assume the management of the hospital, and in 1914 a larger building was erected. It was then able to serve the families of the workers as well.

Single men living in the camp needed accommodations that were clean and reasonable. The Hollinger Mine provided them with a place in its bunkhouses.

Families provided a different challenge. Timmins realized that affordable housing was needed and the building of the famous red and green Hollinger homes began. The houses started out as flat-top affairs (not practical in an area that received huge amounts of snow); peaked roofs and two extra rooms upstairs were added on later.

Hollinger houses under construction.

The superintendents and managers of the Hollinger Mine were housed close to the site. These homes are located in the Hill District and were quite large; they helped attract the needed workforce to the area. Many of the houses still exist today.

While Noah Timmins was deeply involved with the Hollinger Mine, his nephew Jules was also involved; he accompanied the first party to the Hollinger in 1910, at the age of 22. Jules later became the mine manager at the Hollinger, inheriting the position from his Uncle Noah in 1936; he remained in the position until 1961. Jules was a fixture in the early camp and worked underground with the boys in 1914.

Timmins may have been growing, but South Porcupine remained an important town that grew up along the Dome Mine. Many framed and brick buildings were built along Bruce Avenue, the heart of the business district. Sam Bucovetsky opened a store here, and the Toronto Dominion Bank still occupies the same building today.

Golden Avenue led from the business district of South Porcupine to the Dome Mine. Uly Levinson kept the family store on this street until he sold the business in the late 1990s. Boarding houses and the school along with a few taverns could be found on the street.

The Levinsons were pioneers in the camp from the very beginning. Uly, the youngest son, would carry on the family business and continued to tend the store when he was well into his nineties; Minerva would go on to teach in the camp and assist with the founding of the Porcupine Camp Historical Society. The Levinson store is still a fixture on Golden Avenue, and their home still fronts on Bruce Avenue.

The Porcupine became a destination for travellers looking to experience life in a boomtown. Many people were coming up from Toronto on the T&NO railway, while "money men" from New York and Chicago came to investigate their investments, or speculate on new ones. Hotels like the King George in South Porcupine were very popular.

The King George kept a fine dining room; these young ladies were part of the staff complement in the early 1910s. They would have been wearing their Sunday finest for the photograph.

Near the South Porcupine train station, the Connaught Hotel stood in direct competition to the Goldfield Hotel in Timmins. It boasted a fine dining room and well-appointed rooms. Many of the Dome's investors stayed at the Connaught. The hotel burned to the ground in 1916, strangely enough after all of the furnishings had been safely removed "for cleaning and repair."

The well-appointed rotunda of the Connaught Hotel, located near the train station in South Porcupine, was home to many travellers. It included a fireplace and leather chairs for the guests; notice the line of spittoons near the chairs.

Streets in South Porcupine were slowly being straightened and leveled. The roadways were still packed dirt over logs, and appeared to be passable only in the center, but they were miles ahead of the stump-infested bogs of the early camp. Boardwalks near the stores allowed the ladies of the town a clear, dirt-free passage to the merchants.

Timmins was not standing still either; Pine Street was quickly becoming a core feature of the business district, along with Third Avenue. Streets were very passable and wide; horse-drawn carts and pedestrians could easily maneuver around the community. It would only be a short time before automobiles would make their way into the existing traffic.

There is some dispute over who brought the first automobile into the camp. Suffice it to say that while it could not be driven into the camp, an automobile could be brought in by train and driven through the bumpy streets. This gentleman's beauty was evidently the talk of the town.

These two lucky men clear the snow from the front of the Customs Office in South Porcupine, after a particularly brutal winter assault. Snow was not removed from the streets, but banked up along the sidewalks and roadways. While cars would never be able to parade down the streets, horse-drawn sleds were the transportation of choice.

While horses were used to pull sleighs, large dogs could also do the job. This couple glides along the frozen roadway with the assistance of their fit dog team (and straggler).

Sleigh parties were popular winter diversions in the Porcupine. Groups of people would sled to Timmins or South Porcupine for the day. Photographers were eager to shoot the spectacle, as this shooter caught two other camera buffs taking pictures from the telephone pole.

All of that snow was wonderful in the winter, but treacherous in the spring. A quick breakup and thaw could inundate the community quickly. Porcupine Lake overspills its banks in this photograph of Golden Avenue in 1913; the street was impassable but the raised boardwalks were still dry.

Along Evans Street to the train station, the floods were usually at their worst. These homes were quickly pulled from their pilings; canoes, like the one in the background of this picture, were definitely needed. Floods became as threatening to the camp as forest fires.

Mine management realized quickly that a "happy miner was a productive miner" and proceeded to build recreational facilities for their workers and families. As well, if the men were occupied with family pursuits at the halls, they were not meeting in the union halls as frequently. The Dome put in a recreation hall that was used for dances, parties and dinners.

Many of the new immigrants who flocked to the camps built their own halls along ethnic and political lines. This very popular Finnish hall was located in South Porcupine; it was a home away from home for many people who were getting used to their new surroundings, language and way of life.

Many of the halls had stage areas that allowed for speakers, theatre and musical performances. This unidentified hall was like many throughout the camp; the meeting space upstairs was complimented by large kitchens in the basement, where the women would prepare foods like pierogies, cabbage rolls, lasagna, and Finnish coffee bread.

Leo Macsioli was a popular businessman in the community. He owned, besides his car dealership and hotels, many of the camp's theatres. This theatre located at the corner of Third Avenue and Balsam Street showed moving picture films and featured many vaudeville acts from across Canada and the United States.

The Empire Theatre was the first theatre in the Porcupine, dating to 1910. The signs are not exactly politically correct, but the shows were always a large draw.

This photograph of the Timmins Fire Department dates to 1914; many of the crews depended on volunteer forces to shore up their numbers. Increased fire activity around the new towns in 1916 and the devastating blazes of that year encouraged the council of the day to create a brigade of professional fire fighters in Timmins.

Sports and the development of the area are intimately linked; the mines recruited young men who could play hockey, baseball, rugby and individual sports, such as track and field (and of course who would also be good miners and staff). These young men are enjoying a game on an outdoor rink.

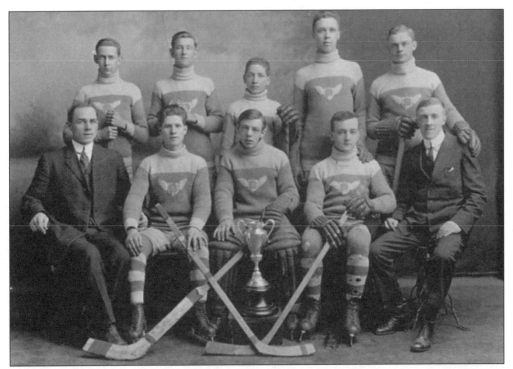

Junior hockey was an important part of life in the north. The towns from Cobalt to Timmins had local teams that played regularly for trophies and bragging rights. Timmins, South Porcupine and Schumacher produced over 40 National Hockey League players from the early 1920s to today.

While many women chose to watch the men from the bleachers, many young girls participated in organized teams. This group of ladies, under the guidance of Coach Marshall, won their tournament. The badge suggests they were from the Porcupine (Porcupine Flyers, perhaps?).

Summer sports were just as popular; the large crowd is enjoying a baseball game on July 1st, 1912. The location of the diamond is somewhere in Timmins. The track and field grounds were located on property owned by Jack Dalton and this was also a very popular spectator sport.

Beer and spirits were readily available in the camp (as seen by the number of saloons and taverns in operation). This tavern also served ice cream and soft drinks. These were popular spots for a drink after a shift, and the tradition continues today.

With families came the need for proper schools and the Porcupine was soon accommodating children in both English and French language classrooms. These children pose in front of their one room schoolhouse, along with their teacher and a young man who could have been the school superintendent.

These young children (properly segregated for the photograph) prepare to celebrate Arbor Day. This photograph dates from 1912 and shows the first true school building in Timmins.

At long last peace was declared and the residents of the area could celebrate. After the war, the camp was inundated with young men returning to look for work. Many Eastern European immigrants made the Porcupine region their home; most came to work in the mines and start a new life in Canada.

Anticipation built up over months when it was announced that His Majesty the Prince of Wales would visit the Porcupine camp on his postwar tour of Canada. Crowds gathered at South Porcupine, but His Majesty's train did not stop; it did stop in Timmins, however, where he visited the Hollinger Mine. The entire event was over in less than a few hours, but crowds were satisfied with the trip.

This group of miners is working underground at the Hollinger. A candle that was stuck in their caps provided light; other candles lined the stopes and work areas. The men did not wear anything but a cloth cap on their heads; they did not have any eye or ear protection.

This miner stands near an area that was recently blasted to reveal a rich vein of gold. Miners worked an eight-hour shift on average, with overtime tacked on when necessary. After the blast, the area would be scaled to remove any loose rock, and then the ore would be moved to the surface for processing.

By 1917, the Hollinger Mine was in full production. The mine would continue to be the main employer in the community until 1968, when it would shut its doors. Exploration continues to this day in the Porcupine Goldfields, with gold and base metal discoveries being made every year.

The early prospectors and adventurers who opened up the Porcupine Goldfields had no idea what awaited them in 1909, the difficult trail often led to a change of heart and a hasty retreat. Those who stuck it out, however, played a major role in the development of Canadian mining history.

Clary Dixon, far right, would continue to chase that elusive find. In 1909, he worked on the Dome claims that would yield some of the world's most spectacular gold specimens. Placer Dome (today part of the Porcupine Venture Group) is still a working gold mine.

The Porcupine Goldfields were quick to attract prospectors who worked the Klondike fields at the turn of the century. Businesses set up shop early in 1911 to take advantage of the traditional boom-bust cycle of mining communities. The camp proved to be more boom than bust.

Even after the fire of 1911, Noah Timmins saw opportunities. At the lot sale, many potential businessmen, speculators, and gamblers came out to pick out a piece of land in the cleared-out bush.

Harry Peters provided a photographic record of life in the early camp that remains invaluable to us today. This photograph features a unique view of Golden City, the main community in the camp at the time. This Sunday afternoon pastime remains popular even today.

The Skys had a number of fine shops in both South Porcupine and Timmins; the family remained in the camp until the mid 1920s. They assisted with the building of the synagogue in Timmins.

A group of miners pose before their shift at one of the many operating mines in
Timmins.

The Pearl Lake Mine began production on the shores of the lake in 1912. It was part of the original McIntyre Claim and became a part of the McIntyre Mine. In the 1960s copper as well as gold were extracted from the site.

Located south of the Hollinger Claims, the Porcupine Crown was originally staked by Hollinger and Gillies for their friend Barney McEnaney. McEnaney sold the claim to mining promoter Jack Hammell, for $200,000 plus one-third stock to be issued. Official records show that this was the first claim in the district that became a mine.

The Porcupine was littered with small mining properties by 1911. Over 120 were estimated to be in some stage of operation. The Rea Gold Mine was one of those companies. It eventually amalgamated with the Coniaurum Mines in Tisdale Township.

Located one kilometre north of Porcupine Lake, this property was the original discovery made by George Bannerman on July 31, 1909. By 1911, a shaft was sunk and a mill was set up to handle the ore.

The land was staked in 1909 by Joe Vipond and Bill Davidson and became one of the earliest mines in the area. In 1922, the Vipond and the porcupine Crown merged to become the Vipond Consolidated Mine, which later became part of the Hollinger Consolidated Mine. The mine continued to operate until 1941 and produced 414,367 ounces of gold.

The West Dome Mine near the Dome Property became a part of the Paymaster Mine in 1915. In 1911, it was the site of the tragic deaths of the mine manager and his family during the Porcupine Fire of 1911.

Located east of Porcupine Lake, the Hunter Mine was staked early in the Gold Rush (1907) for A.G. Hunter, a Toronto lawyer. The mine was worked at intervals until 1914, when it closed due to the First World War. It reopened briefly in the 1930s under the name Porcupine Lake Mine. In the 1980s, it was again reworked under the name of Wabigoon Resources.

Located near the McIntyre Mine, the Jupiter Mine was opened in 1911. It quickly became a part of the larger mine.

Located near the Hollinger Mine, eight claims were staked in 1909. The mine continued to produce until 1943; during that time, 149,340 ounces of gold were extracted from the ore, as well as 22,045 ounces of silver.

Part of the area near the Big Dome Mine, this site sank two shafts in 1912. In this photo, the two headframes are visible along with the water tower and mill—all important features of the property.

The original property was staked by the Wilson/Preston team in 1909. The mine was incorporated in 1911, but did not start production until 1939. Once it started work, however, the mine produced well over $50 million worth of gold.

One of the earliest mines in the area, the Foley O'Brien started to produce in 1911. In 1912, a fire burned the mine buildings to the ground while the manager played baseball to avoid losing a $600 bet. Dome Mines subsequently purchased the property.

This photograph shows a rough headframe and conveyor belt that was part of the Triple Lake Mine. No records of production exist for the site.

The Schumacher Mine located near Pearl Lake was a large operation in 1916.

Two prospectors work a windlass at the Two-And-One Mine in Whitney Township. The miners descended into the shaft using ladders while one or two other men worked the bucket that would be filled with raw ore. The Two-And-One Mine was typical of many of the small operations in the area: it never produced much gold, but it played an important part in the history of the Porcupine Goldfields.

Harry Peters was an early photographer in the Porcupine Camp. His work helped document the development of the Porcupine Goldfields and has proven to be an invaluable record of the past. A large part of his photograph collection can be found in the archives of the Timmins Museum.

End Notes

Remnants of the mines are still visible, for the most part. The Hollinger buildings still dominate the skyline in Timmins. The McIntyre Mine's huge headframe sits on Pearl Lake, a testimony to years of production at the 8000-foot levels. The Dome is still a working mine, digging up the land around the Ida Maud and the Golden Staircase.

The "City with a Heart of Gold" may have grown, but it remains reliant on the resource-based industries that were a part of her beginning: mining, lumbering and outfitting. The communities that make up the Porcupine Goldfields have played an important part in the mining history of Canada. It will be up to the descendants of those first pioneers to make sure that the heritage of northern Ontario is not lost or ignored, but celebrated and preserved.

Bibliography

Barnes, Michael. *Killer in the Bush*. Erin, ON: Boston Mills Press,
 1987.

_____. *Eighteenth Annual report of the Bureau of Mines, 1909*.
 Toronto, ON: L.K. Cameron, Printer to the King's Most Excellent Majesty,
 1909.

Caesar, Cliff. *The Prospector*. Timmins, ON: Porcupine Prospector's
 Association, 1939.

_____. *Information about the Hollinger Gold Mines at Timmins*
 booklet issued to the Canadian Press Association, June 6th, 1913.

_____. *A Souvenir from the Municipality*, Timmins Board
 of Trade, for the Canadian Press Association, June 6th, 1913.

Hoffman, Arnold. *Free Gold: The Story of Canadian Mining*. New York, NY:
 Associated Book Service, 1947.

_____. *Nineteenth Annual report of the Bureau of Mines, 1910.*
 Toronto, ON: L.K. Cameron, Printer to the King's Most Excellent
 Majesty, 1910.

Smith, Philip. *Harvest from the Rock*. Toronto, ON: Macmillan of Canada,
 1986.

Surtees, Robert. *The Northern Connection: Ontario Northland Since 1902*.
 Toronto, ON: Captus Press, 1992.

Tremblay, Rodolphe. *Timmins, métropole de l'or*. Sudbury, ON: La société
 Historique du nouvel-Ontario, 1951.

Tucker, Albert. *Steam Into Wilderness*. Toronto, ON: Fitzhenry and Whiteside,
 1978.

Oral history accounts from the Porcupine Fire of 1911, archives of the Timmins
 Museum: NEC.

Exhibition notes and curatorial notes, 1975-1981, Timmins Museum: NEC.

The Porcupine Advance. Timmins. 1912, 1915, 1916, 1917, 1918.

The Cobalt Nugget. Cobalt. 1910,1911.

All of the photos in this book come from the archives of the Timmins Museum:
 National Exhibition Center.